MATHadazzles

Mind Stretch Puzzles

Reasoning with Numbers Volume 2

Authors

Carole Greenes

Mary Cavanagh

Contributors

Renee Ashlock, Channing Bogle, Nancy Foote, Kristi Larson,
Katie MacDonald, Holly Natonick, Jane Placencia,
Theresa Thivierge, Jen Tom, Diana Volanti, Cara Wright, Michael Xu

Editors

Javy Duarte, Senior Editor

James Kim

Jason Luc

Lesley Le

Shelley Tingey

Tanner Wolfram

Cover Design

Yifan Tian

MATHadazzles *Mind Stretch Puzzles*
Reasoning with Numbers Volume 2

MATHadazzles are number puzzles that will develop your logical reasoning abilities, your sense of numbers (different types of numbers, their characteristics, and operations with them), and your persistence in solving problems. Once you start, you won't be able to stop UNTIL you successfully solve all of the puzzles!

What is a MATHadazzle? A MATHadazzle is a 3-by-3 grid with circles at the end of each row and column. Some grid cells have clues about the numbers that will fill those cells. Circles contain some of the sums of the numbers in that row or column.

What's Your Job? Based on clues provided in some of the grid cells, you place the numbers 1 - 9 in the nine cells AND the given sums in the empty circles at the ends of the rows and columns. Sounds challenging? It is, but don't give up.

What are The Clues? Clues describe types of numbers, characteristics of those numbers, or results of operations with the numbers. There are 15 types of clues. These may appear singly or in combination.

> **Even:** Number that is divisible by 2. In the set of numbers $1 - 9$, the even numbers are 0, 2, 4, 6 and 8.

> **Odd:** Number that is not even. In the set of numbers 1-9, the odd numbers are 1, 3, 5 7, and 9.

> **Multiple:** Number that is the product of a counting number and a whole number. Examples: 8 is a multiple of 4, because 2 x 4 = 8, and 6 is a multiple of 1 because 6 x 1 = 6.

> **Factor**: One of two or more numbers that produce a product. Examples: 2 and 3 are factors of 6 because 2 x 3 = 6, and 2 and 4 are factors of 8, because 2 x 4 = 8.

> **Prime:** Number divisible by only two numbers, itself and 1. Important: Itself and 1 must be different numbers. Examples: 3 is only divisible by 1 and 3. 7 is only divisible by 1 and 7.

> **Composite:** Number greater than 1 that is not a prime number. That is, it has more than two different factors. Example: 4 can be divided by 1, 2 and 4. 8 can be divided by 1, 2, 4 and 8.

> **Square Number:** Number that is the product of a non-zero number and itself. Examples: 4 is a square number because 2 x 2 = 4, and 9 is a square number because 3 x 3 = 9.

Square Root of a Number: Number that when multiplied by itself has a product equal to a given number. The square root of 16 is 4, because 4 x 4 = 16. The square root symbol is $\sqrt{}$. Example: $\sqrt{16} = 4$

Cubic Number: Number that is the product of three same non-zero numbers. Examples: 8 is a cubic number because 2 x 2 x 2 = 8. 1 is cubic number because 1 x 1 x 1 = 1.

Triangular Number: Number that is the sum of consecutive counting numbers, beginning with 1. Examples: 3 is a triangular number because 1 + 2 = 3, and 6 is a triangular number because 1 + 2 + 3= 6.

Perfect Number: Number whose proper factors, when added, are equal to itself. Example: 6 is a perfect number. Its proper factors (factors not equal to itself) are 1, 2, and 3, and 1 + 2 + 3 = 6.

Abundant Number: Number whose proper factors add to a number greater than itself. Example: 12 is an abundant number because 1 + 2 + 3 + 4 + 6 = 16 and 16 > 12.

Deficient Number: Number whose proper factors add to a number less than itself. Example: 4 is a deficient number because 1 + 2 = 3 and 3 < 4.

Exponent: A small number to the right and above a base number that indicates the number of times that base number should be used as a factor in multiplication. Example: 2^3 = 2 x 2 x 2 = 8. In this example, 2 is the base number, 3 is the exponent, and 8 is the product. The product 8 is also referred to as a **power of 2.**

Computation Operations: All four operations with whole numbers are incorporated into the clues. These are: addition, subtraction, multiplication and division. When several different operations appear in a number sentence, then the order in which they are performed follow the Fundamental Order of Operations: Parentheses (all operations within parentheses are computed first), Powers (all base numbers with exponents), Multiplication and Division (from left to right), and finally, Addition and Subtraction (from left to right).

In Volume II, you will find 78 MATHadazzles with answers. If you are interested in improving your dazzling solution talents, consider getting Volume I, with more puzzles like those in Volume II. Volume III uses the same format, but this time with integers.

Enjoy Solving!

Your MATHadazzling Authors, Contributors and Editors

1

Put these numbers in the squares   1   2   3   4   5   6   7   8   9

Add across →

Add down ↓

Sums are in ◯

Put these numbers in the circles: 8, 11, 13

Multiple of 3^2		
		Prime
Multiple of 3	Multiple of 3	

◯ 21

◯

◯

◯ 20   ◯   ◯ 17

Put these numbers in the squares   1   2   3   4   5   6   7   8   9

Add across ⟶

Add down ↓

Sums are in ◯

Put these numbers in the circles: 13, 15, 18

64 ÷ 8		72 ÷ 8
Factor of 4		Factor of 4

21

◯

9

14   ◯   ◯

3

Put these numbers in the squares 1 2 3 4 5 6 7 8 9

Add across →

Add down ↓

Sums are in ◯

Put these numbers in the circles: 13, 16, 18

24 ÷ 3	Even	Even	◯
Multiple of 5			◯
		18 ÷ 6	**14**
15	◯	**14**	

Put these numbers in the squares   1   2   3   4   5   6   7   8   9

Add across   →

Add down   ↓

Sums are in   ◯

Put these numbers in the circles: 14, 15, 18

$96 \div 16 + 2$	Even	Multiple of 3	**21**
	Odd		◯
Even factor of 24	Odd	$121 - 120$	**6**
◯	◯	**16**	

Put these numbers in the squares   1   2   3   4   5   6   7   8   9

Add across ⟶

Add down ↓

Sums are in ◯

Put these numbers in the circles: 8, 10, 14

$123 - 116$		$2 \times 2 \div 2$
	$96 \div 16$	
Even		Odd

◯ **18**

◯

◯ **17**

◯   ◯ **23**   ◯

Put these numbers in the squares 1 2 3 4 5 6 7 8 9

Add across ⟶

Add down ↓

Sums are in ◯

Put these numbers in the circles: 13, 13, 13

$108 \div 12$			**19**
Factor of 24		Odd	◯
	Odd	$63 \div 9$	◯
16	◯	**16**	

Put these numbers in the squares 1 2 3 4 5 6 7 8 9

Add across ⟶

Add down ↓

Sums are in ◯

Put these numbers in the circles: 13, 14, 15, 18

Even	$36 \div 9$	$21 \div 21$	◯
$17 - 14$	Multiple of 3		◯
Odd	2×1		◯

(**16**)   (   )   (**14**)

Put these numbers in the squares   1   2   3   4   5   6   7   8   9

Add across $\longrightarrow$

Add down $\downarrow$

Sums are in   ◯

Put these numbers in the circles: 9, 12, 15

$36 \div 9 - 1$	$1 \times 1 \div 1 \times 1$		◯
$117 \div 13$		$100 \div 20 + 2$	**24**
	Even		◯
16	◯	**14**	

Put these numbers in the squares   1   2   3   4   5   6   7   8   9

Add across $\longrightarrow$

Add down $\downarrow$

Sums are in ◯

Put these numbers in the circles: 7, 15, 20

$13 - 11$		Even	◯
$25 \div 5$		$41 - 32$	◯
	4×2		**18**
14	◯	**16**	

Put these numbers in the squares   1   2   3   4   5   6   7   8   9

Add across ⟶

Add down ↓

Sums are in ◯

Put these numbers in the circles: 14, 16, 20

	$1 + 2 \times 3$		**11**
$18 - (3 \times 3)$	$(8 + 2) \div 2$		◯
		$2 \times 2 \times 2$	◯
12	◯	**17**	

Put these numbers in the squares   1   2   3   4   5   6   7   8   9

Add across ⟶

Add down ↓

Sums are in ◯

Put these numbers in the circles: 8, 15, 15, 19

$6 \div (3 \times 2)$			**11**
Factor of 10	$1 + 1 \times 5$	Square	◯
Even Prime			◯
◯	**22**	◯	

12

Put these numbers in the squares   1   2   3   4   5   6   7   8   9

Add across ⟶

Add down ↓

Sums are in ◯

Put these numbers in the circles: 13, 15, 17

	$24 \div 4 + 1$	Square	◯
	Square	Square	**15**
Multiple of 1 and 2		Multiple of 1, 2, and 3	◯

◯   **11**   **19**

13

Put these numbers in the squares   1   2   3   4   5   6   7   8   9

Add across →

Add down ↓

Sums are in ◯

Put these numbers in the circles: 12, 13, 14, 18

$3 \times (8 - 5)$	$6 \times 3 \div 18$	Prime	◯
		$24 \div 8 \times 2$	◯
	Square		◯

◯ **16**   ◯   ◯ **17**

14

Put these numbers in the squares   1   2   3   4   5   6   7   8   9

Add across ⟶

Add down ↓

Sums are in ◯

Put these numbers in the circles: 16, 17, 18

Factor of 27		Odd	◯
	Even Prime	Square	⑫
	Multiple of 5	Multiple of 4	◯
◯	⑭	⑬	

15

Put these numbers in the squares   1   2   3   4   5   6   7   8   9

Add across ⟶

Add down ↓

Sums are in ◯

Put these numbers in the circles: 13, 14, 20

$72 \div 8 - 2$		Odd Square	◯
$8 \div 8$			(10)
$35 \div 5 - 2$		Even Prime	(15)
◯	(18)	◯	

16

Put these numbers in the squares   1   2   3   4   5   6   7   8   9

Add across   →

Add down   ↓

Sums are in   ◯

Put these numbers in the circles: 15, 17, 18

	Prime	Square	
			9
Square	$42 \times 3 \div 18$		**21**
		Composite	◯

◯   **10**   ◯

Put these numbers in the squares   1   2   3   4   5   6   7   8   9

Add across ⟶

Add down ↓

Sums are in ◯

Put these numbers in the circles: 11, 15, 16

Odd less than 2		Odd Factor of 10	$\bigcirc$ **14**
Multiple of 2 and 3	$10 \div (2 + 3)$		$\bigcirc$ **15**
Even Square			$\bigcirc$
$\bigcirc$	$\bigcirc$ **19**	$\bigcirc$	

Put these numbers in the squares 1 2 3 4 5 6 7 8 9

Add across ⟶

Add down ↓

Sums are in ◯

Put these numbers in the circles: 6, 15, 22

$13 - 6$		Square
$25 \div 5^2$		
	$10 \div 2$	$(37 + 8) \div 5$

◯ 17

◯

◯

◯ 16   ◯ 14   ◯

Put these numbers in the squares   1   2   3   4   5   6   7   8   9

Add across ⟶

Add down ↓

Sums are in ◯

Put these numbers in the circles: 6, 16, 18

$10^0 \times 10^0$		Triangular
$2^2 + 5$		Perfect

Circles (right): ◯, **20**, **19**

Circles (bottom): ◯, **11**, ◯

Put these numbers in the squares   1   2   3   4   5   6   7   8   9

Add across →

Add down ↓

Sums are in ◯

Put these numbers in the circles: 14, 18, 19

Even	Perfect		**13**
$\sqrt{9} \times \sqrt{9}$		$\sqrt{9} + 9^0$	**14**
Composite		Triangular	◯
◯	◯	**12**	

Put these numbers in the squares   1   2   3   4   5   6   7   8   9

Add across ⟶

Add down ↓

Sums are in ◯

Put these numbers in the circles: 12, 15, 18

		Factor of 12	
			6
$(2+8) \div 2$		$\sqrt{81} - \sqrt{9}$	
			15
	$\sqrt{49}$	Odd	
			24

◯   ◯   ◯

Put these numbers in the squares   1   2   3   4   5   6   7   8   9

Add across $\longrightarrow$

Add down $\downarrow$

Sums are in ◯

Put these numbers in the circles: 12, 12, 17

$\sqrt{25}$		Perfect	◯
			◯
	Square	Square	**21**
15	◯	**13**	

Put these numbers in the squares   1   2   3   4   5   6   7   8   9

Add across   →

Add down   ↓

Sums are in   ◯

Put these numbers in the circles: 12, 16, 17

		$\sqrt[3]{27}$
$72 \div 8$		
Power of 4	Power of 4	

Circles on the right (top to bottom): ◯ ◯ ◯

Bottom circles: **18**   **12**   **15**

Put these numbers in the squares   1   2   3   4   5   6   7   8   9

Add across ⟶

Add down ↓

Sums are in ◯

Put these numbers in the circles: 16, 20, 20

$\sqrt{64} \times 2 \div 4$	Square		
	$2^3 - 2$	$\sqrt{49}$	**15**
		Multiple of 2	◯
9	◯	◯	

(circle top right: **10**)

Put these numbers in the squares   1   2   3   4   5   6   7   8   9

Add across ⟶

Add down ↓

Sums are in ◯

Put these numbers in the circles: 6, 14, 15

$(64 + 9) \div 73$	Composite	Power of 2
	$\sqrt{121} - \sqrt{9}$	
$2^0 + 2^1$		Perfect

◯

◯

16

◯   **24**   **15**

Put these numbers in the squares   1   2   3   4   5   6   7   8   9

Add across →

Add down ↓

Sums are in ◯

Put these numbers in the circles: 12, 12, 15

Odd		$\sqrt{16}$	◯
Multiple of 2	Prime		**16**
$\sqrt{225} \div 3$		$72 \div 3 \div 8$	**17**
◯	**18**	◯	

Put these numbers in the squares   1   2   3   4   5   6   7   8   9

Add across →

Add down ↓

Sums are in ◯

Put these numbers in the circles: 11, 11, 17, 19

2^3			◯
	$\sqrt{16} + 8^0$	Odd	**15**
Even		$(10 - 3)^0$	◯
17	◯	◯	

Put these numbers in the squares   1   2   3   4   5   6   7   8   9

Add across ➜

Add down ↓

Sums are in ◯

Put these numbers in the circles: 9, 12, 15

	$18 - 5 \times 3$		◯
$13 - 2 \times 4$	Factor of 12		**18**
$2 \times 2 \times 2$		$13 - 6 \times 2$	**18**
◯	**18**	◯	

Put these numbers in the squares   1   2   3   4   5   6   7   8   9

Add across ⟶

Add down ↓

Sums are in ◯

Put these numbers in the circles: 12, 18, 18

$50 \div 5 \div 2$	$9 - 3$		◯
	Factor of 9	Odd	◯
	Factor of 9	$10 \div 5 + 2$	⑨
⑮	⑱	◯	

30

Put these numbers in the squares   1   2   3   4   5   6   7   8   9

Add across $\longrightarrow$

Add down $\downarrow$

Sums are in ◯

Put these numbers in the circles: 15, 18, 21

	Odd	Factor of 8	
			12
$1 + 2 + 0$	2×2		**12**
		$32 \div 4$	◯
◯	**12**	◯	

31

Put these numbers in the squares   1   2   3   4   5   6   7   8   9

Add across ⟶

Add down ↓

Sums are in ◯

Put these numbers in the circles: 11, 12, 16, 18

Odd		$6 - 4 \div 2 \times 2$
$\sqrt{25}$	$9 - 6 \div 3$	Perfect
Odd	Odd	

◯

◯

◯

(**9**)   (**24**)   ( )

Put these numbers in the squares   1   2   3   4   5   6   7   8   9

Add across ⟶

Add down ↓

Sums are in ◯

Put these numbers in the circles: 13, 13, 16

Even		Even
	$9 - (3 \times 3) + 1$	$9 - (3 \div 3) + 1$
$6 - (4 \div 2) + 1$	$8 - 2 + (4 \div 2)$	Odd

Circles (right side): ◯, **16**, **16**

Circles (bottom): ◯, ◯, **16**

Put these numbers in the squares   1   2   3   4   5   6   7   8   9

Add across $\longrightarrow$

Add down $\downarrow$

Sums are in $\bigcirc$

Put these numbers in the circles: 11, 18, 18

$2 \times 2 \times 2$	$10 - 6$	
		$27 \div 9$
Factor of 14 and 21		$5 \times 2 - 8$

$\bigcirc$ **18**

$\bigcirc$ **9**

$\bigcirc$

$\bigcirc$ **16** $\bigcirc$ $\bigcirc$

Put these numbers in the squares 1 2 3 4 5 6 7 8 9

Add across →

Add down ↓

Sums are in ◯

Put these numbers in the circles: 13, 15, 16

$25 \div 5$	Odd		◯
		$3 \times 3 - 1$	**14**
3×3	$5 \times 2 - 4$		◯
16	◯	**16**	

Put these numbers in the squares   1   2   3   4   5   6   7   8   9

Add across ⟶

Add down ↓

Sums are in ◯

Put these numbers in the circles: 12, 14, 14

Factor of 16		$7 \div 7$
Odd factor of 6	Even	$27 \div 3$

Circles: **13**, ( ), **18**, ( ), **19**, ( )

Put these numbers in the squares   1   2   3   4   5   6   7   8   9

Add across →

Add down ↓

Sums are in ◯

Put these numbers in the circles: 12, 16, 16

Odd	$6 \div 2 + 4$	
	1×1	Multiple of 3
Factor 10		$10 - 6$

Circles (right side): **12**, ◯, **17**

Circles (bottom): **17**, ◯, ◯

37

Put these numbers in the squares   1   2   3   4   5   6   7   8   9

Add across   →

Add down   ↓

Sums are in ◯

Put these numbers in the circles: 9, 11, 18

			◯
	$21 \div 3$		**14**
	Multiple of 4	$35 \div 7$	**22**
◯	**16**	◯	

38

Put these numbers in the squares   1   2   3   4   5   6   7   8   9

Add across ⟶

Add down ↓

Sums are in ◯

Put these numbers in the circles: 12, 17, 18, 22

$5^2 \div 5$			(10)
Factor of 16	$1 + 1 \times 5$	Prime	◯
Odd Square			◯

◯   ◯   (11)

39

Put these numbers in the squares   1   2   3   4   5   6   7   8   9

Add across →

Add down ↓

Sums are in ◯

Put these numbers in the circles: 13, 14, 18

Prime factor of 49	Even prime		◯
Neither prime nor composite		$(7-4) \times 3$	◯
			14

◯   13   18

40

Put these numbers in the squares   1   2   3   4   5   6   7   8   9

Add across →

Add down ↓

Sums are in ◯

Put these numbers in the circles: 16, 20, 22

Even square		Multiple of 9	
			◯
Square	$5 \times 9 \div 15$		⑨
			◯

⑦   ⑯   ◯

41

Put these numbers in the squares   1   2   3   4   5   6   7   8   9

Add across ⟶

Add down ↓

Sums are in ◯

Put these numbers in the circles: 11, 15, 19

Multiple of 5	Odd prime	Odd prime
64 ÷ 8		Even Prime

◯ ◯ ◯

(14)   (16)   (15)

Put these numbers in the squares   1   2   3   4   5   6   7   8   9

Add across ⟶

Add down ↓

Sums are in ◯

Put these numbers in the circles: 8, 9, 20, 20

$6 - 5 \div (2 + 3)$		Odd
	Factor of 5	
Square	Factor of 18	

◯ **16**

◯

◯

◯   ◯   ◯ **17**

Put these numbers in the squares 1   2   3   4   5   6   7   8   9

Add across →

Add down ↓

Sums are in ◯

Put these numbers in the circles: 13, 13, 21, 22

Square	Prime	
$(9 + 9) \div 2$		$64 \div 2 \div 4$
Prime	Square	

Circles: **10**, **11**

44

Put these numbers in the squares   1   2   3   4   5   6   7   8   9

Add across ⟶

Add down ↓

Sums are in ◯

Put these numbers in the circles: 12, 16, 17, 20

Even		Square
Multiple of 2		$24 - 9 \times 2$
Prime	Odd	

Circles: 9, 16

45

Put these numbers in the squares   1   2   3   4   5   6   7   8   9

Add across →

Add down ↓

Sums are in ◯

Put these numbers in the circles: 11, 13, 14, 19

$2^2 + 2^1$		Square	◯
	Square	Prime	**21**
Odd		$5 \times (4 - 2) - 2$	◯
12	◯	◯	

Put these numbers in the squares   1   2   3   4   5   6   7   8   9

Add across →

Add down ↓

Sums are in ◯

Put these numbers in the circles: 7, 9, 17

$2^1 \times 3^1$		3^2
		Even
	Triangular	

◯ **20**

◯

◯ **18**

◯   ◯   ◯ **19**

Put these numbers in the squares   1   2   3   4   5   6   7   8   9

Add across →

Add down ↓

Sums are in ◯

Put these numbers in the circles: 9, 16, 24

		$\sqrt{25}$	◯
	Perfect	Even square	**12**
3^2			◯
12	◯	**17**	

48

Put these numbers in the squares   1   2   3   4   5   6   7   8   9

Add across ➝

Add down ↓

Sums are in ◯

Put these numbers in the circles: 12, 12, 14, 15

		$\sqrt{16}$
Perfect		Even prime factor of 42

18

◯

◯

◯   **19**   ◯

49

Put these numbers in the squares   1   2   3   4   5   6   7   8   9

Add across ⟶

Add down ↓

Sums are in ◯

Put these numbers in the circles: 11, 14, 15, 18

	$2^3 - 5$	$4^3 \div 8$	**16**
	$\sqrt{49}$	Perfect	◯
			◯
16	◯	◯	

50

Put these numbers in the squares   1   2   3   4   5   6   7   8   9

Add across →

Add down ↓

Sums are in ◯

Put these numbers in the circles: 14, 17, 19, 19

		$\sqrt{25} + \sqrt{16}$	◯
Even prime	6^0		**9**
$\sqrt{81} - \sqrt{4}$			◯
◯	**12**	◯	

51

Put these numbers in the squares   1   2   3   4   5   6   7   8   9

Add across ⟶

Add down ↓

Sums are in ◯

Put these numbers in the circles: 9, 15, 23

Odd			◯
	Even prime	Factor of 12 and 16	**13**
	$4 \times \sqrt{4}$		◯

◯ **19**   ◯ **11**   ◯

Put these numbers in the squares   **1 2 3 4 5 6 7 8 9**

Add across $\longrightarrow$

Add down $\downarrow$

Sums are in ◯

Put these numbers in the circles: 9, 12, 15

	Perfect		
			◯
	Triangular		**18**
Even square			**18**

◯   **18**   ◯

Put these numbers in the squares   1   2   3   4   5   6   7   8   9

Add across ⟶

Add down ↓

Sums are in ◯

Put these numbers in the circles: 16, 17, 21

	Perfect	$7^0 \times 7^1$	**18**
Even square	3^2		◯
Triangular	Even Prime		**6**
12	◯	◯	

54

Put these numbers in the squares   1   2   3   4   5   6   7   8   9

Add across $\longrightarrow$

Add down $\downarrow$

Sums are in $\bigcirc$

Put these numbers in the circles: 7, 14, 24

	Perfect		$\left(14\right)$
	10^0	$\sqrt[3]{8}$	$\bigcirc$
Square			$\bigcirc$

$\left(16\right)$ $\left(15\right)$ $\bigcirc$

55

Put these numbers in the squares   1   2   3   4   5   6   7   8   9

Add across →

Add down ↓

Sums are in ◯

Put these numbers in the circles: 12, 16, 17

	Factor of 9	Even	
			⬤ 12
	$51 \div 17$		⬤ 16
		$32 \div 8$	⬤ 17

◯   ◯   ◯

Put these numbers in the squares   1   2   3   4   5   6   7   8   9

Add across $\longrightarrow$

Add down $\downarrow$

Sums are in $\bigcirc$

Put these numbers in the circles: 12, 15, 18

			21
$27 \div 3$	Factor of all numbers		**12**
Factor of 9	$2 \times 2 + 0$	Odd	$\bigcirc$

    $\bigcirc$ **12** $\bigcirc$

Put these numbers in the squares   1   2   3   4   5   6   7   8   9

Add across ⟶

Add down ↓

Sums are in ◯

Put these numbers in the circles: 17, 17, 18

$64 \div 2^3$	Odd	Multiple of 3	◯
Odd	Odd		**10**
Odd	$10 - 7 + 1$	Multiple of 3	◯
18	**10**	◯	

Put these numbers in the squares   1   2   3   4   5   6   7   8   9

Add across ⟶

Add down ↓

Sums are in ◯

Put these numbers in the circles: 12, 17, 18

Multiple of 3		
	$9 - 6 \div 3 \times 2 + 2$	Odd
	Odd	Odd

Circles to the right of grid:
(top) ◯ (middle) **10** (bottom) ◯

Circles below grid:
◯   **24**   **9**

Put these numbers in the squares   1   2   3   4   5   6   7   8   9

Add across ⟶

Add down ↓

Sums are in ◯

Put these numbers in the circles: 14, 15, 16

Odd	Odd	Multiple of 4	
			16
	$8 - 4 \div 2 + 3$	Odd	◯
$9 - 8 \div 2 + 1$	Factor of 6	Odd	◯

◯ **15**   ◯ **14**   ◯

Put these numbers in the squares   1   2   3   4   5   6   7   8   9

Add across   →

Add down   ↓

Sums are in   ◯

Put these numbers in the circles: 10, 12, 15

$7 - 3 \times 2$	Factor of 25		
Multiple of 3	Multiple of 3	$9 - 3 + 2$	**23**
	Odd	Odd	

9   **21**

Put these numbers in the squares   1   2   3   4   5   6   7   8   9

Add across →

Add down ↓

Sums are in ◯

Put these numbers in the circles: 12, 14, 17

		Factor of 14	◯
	$10 \div 2$	Multiple of 3	**18**
$8 \div 2 + 4$			**15**

14   ◯   ◯

Put these numbers in the squares   1   2   3   4   5   6   7   8   9

Add across $\longrightarrow$

Add down $\downarrow$

Sums are in ◯

Put these numbers in the circles: 11, 12, 18

Even		$4 \times 4 \div 2$
$32 \div 8$	Factor of 14	
		$1 + 2 \times 4$

Circles to the right: **16**, ◯, **17**

Circles below: ◯, **16**, ◯

Put these numbers in the squares   1   2   3   4   5   6   7   8   9

Add across $\longrightarrow$

Add down $\downarrow$

Sums are in ◯

Put these numbers in the circles: 11, 14, 18

Factor of 3		Factor of 3
Factor of 6		
	8×8^0	Even Prime

13

22   **12**

64

Put these numbers in the squares   1   2   3   4   5   6   7   8   9

Add across ⟶

Add down ↓

Sums are in ◯

Put these numbers in the circles: 11, 15, 19, 21

Factor of 51		
	Factor of 55	Square
$4 \div 2 \div (1 + 1)$	Even	Even

◯ **9**

◯

◯

◯   ◯ **15**   ◯

Put these numbers in the squares   1   2   3   4   5   6   7   8   9

Add across ⟶

Add down ↓

Sums are in ◯

Put these numbers in the circles: 11, 11, 13, 21

	Square	$21 - 5 \times 3$	**23**
	Prime		◯
$\sqrt{16}$		Even	**11**

◯ ◯ ◯

Put these numbers in the squares   1   2   3   4   5   6   7   8   9

Add across →

Add down ↓

Sums are in ◯

Put these numbers in the circles: 8, 9, 16, 20

Square		
	Square and triangular	Prime and triangular
$14 \times 2 \div 4$		Even

Circles: (right side, top) ◯   (middle) ◯   **17**

Circles: (bottom) **20**   ◯   ◯

67

Put these numbers in the squares   1   2   3   4   5   6   7   8   9

Add across ⟶

Add down ↓

Sums are in ◯

Put these numbers in the circles: 10, 12, 18, 23

	Square	$3 + 4 \div 2 \times 2$
Even		
	Odd	$2 + 3 \times 2$

12

15

68

Put these numbers in the squares   1   2   3   4   5   6   7   8   9

Add across →

Add down ↓

Sums are in ◯

Put these numbers in the circles: 7, 20, 22

		Factor of all numbers	⬤ **9**
Odd		Even square	◯
			⬤ **16**
◯	⬤ **16**	◯	

Put these numbers in the squares   1   2   3   4   5   6   7   8   9

Add across $\longrightarrow$

Add down $\downarrow$

Sums are in ◯

Put these numbers in the circles: 9, 19, 22

		$36 \div (3 + 3)$	◯
Factor of 8		Factor of 8	**12**
	$16 \div 2 - 3$	Factor of 8	**11**
◯	**17**	◯	

Put these numbers in the squares   1   2   3   4   5   6   7   8   9

Add across →

Add down ↓

Sums are in ◯

Put these numbers in the circles: 15, 16, 17

Factor of 36	Factor of 25		
	Odd square	Factor of all numbers	◯
Multiple of 4			**13**

◯   **21**   ◯   **8**

Put these numbers in the squares   1   2   3   4   5   6   7   8   9

Add across ⟶

Add down ↓

Sums are in ◯

Put these numbers in the circles: 13, 14, 22

	Perfect	Odd factor of 8	⟶ **10**
Odd	Odd prime	$(7 + 9) \div 2$	◯
Even			◯

◯   **18**   **13**

72

Put these numbers in the squares   1   2   3   4   5   6   7   8   9

Add across →

Add down ↓

Sums are in ◯

Put these numbers in the circles: 11, 14, 17, 22

		$3 + 3 \times 2$	◯
	$2 \div 1$		⑥
Square			◯
◯	◯	⑳	

73

Put these numbers in the squares   1   2   3   4   5   6   7   8   9

Add across ➡️

Add down ⬇️

Sums are in ◯

Put these numbers in the circles: 14, 14, 19

Even	2^3+1	
Factor of 7		Factor of 16
$\sqrt{25}$		Factor of 7

Circles: **19**, **12**, **12**

Put these numbers in the squares   1   2   3   4   5   6   7   8   9

Add across →

Add down ↓

Sums are in ◯

Put these numbers in the circles: 12, 14, 15

	$\sqrt{36}$	$\sqrt{16}$	◯
			18
$\sqrt{4}$			◯
16	◯	15	

Put these numbers in the squares   1   2   3   4   5   6   7   8   9

Add across →

Add down ↓

Sums are in ◯

Put these numbers in the circles: 15, 17, 18

Perfect		$4^3 \div 16$
Odd square	Even	
		Odd

Circles: **12**, **18**, **10**

76

Put these numbers in the squares   1   2   3   4   5   6   7   8   9

Add across ⟶

Add down ↓

Sums are in ◯

Put these numbers in the circles: 11, 15, 19

Multiple of 3			◯
	Square		⑦
	2^3	Multiple of 3	⑲
◯	⑲	◯	

Put these numbers in the squares   1   2   3   4   5   6   7   8   9

Add across ⟶

Add down ↓

Sums are in ◯

Put these numbers in the circles: 14, 16, 24

	$\sqrt{6^2}$		◯ 14
Even			◯
	Even	$\sqrt{81}$	◯

◯ 6   ◯ 15   ◯

Put these numbers in the squares   1   2   3   4   5   6   7   8   9

Add across   →

Add down   ↓

Sums are in   ◯

Put these numbers in the circles: 8, 10, 15, 18

$\sqrt{81}$			**19**
	Perfect	$5^0 \times 5^1$	◯
$\sqrt{16}$		$\sqrt{9}$	◯
20	◯	◯	

1.

9	4	8	(21)
5	1	7	(13)
6	3	2	(11)
(20)	(8)	(17)	

2.

8	4	9	(21)
5	3	7	(15)
1	6	2	(9)
(14)	(13)	(18)	

3.

8	6	4	(18)
5	1	7	(13)
2	9	3	(14)
(15)	(16)	(14)	

4.

8	4	9	(21)
5	7	6	(18)
2	3	1	(6)
(15)	(14)	(16)	

5.

7	9	2	(18)
3	6	1	(10)
4	8	5	(17)
(14)	(23)	(8)	

6.

9	4	6	(19)
2	8	3	(13)
5	1	7	(13)
(16)	(13)	(16)	

7.

8	4	1	(13)
3	9	6	(18)
5	2	7	(14)
(16)	(15)	(14)	

8.

3	1	5	(9)
9	8	7	(24)
4	6	2	(12)
(16)	(15)	(14)	

9.

2	1	4	(7)
5	6	9	(20)
7	8	3	(18)
(14)	(15)	(16)	

10.

1	7	3	(11)
9	5	6	(20)
2	4	8	(14)
(12)	(16)	(17)	

11.

1	7	3	(11)
5	6	4	(15)
2	9	8	(19)
(8)	(22)	(15)	

12.

2	7	4	(13)
5	1	9	(15)
8	3	6	(17)
(15)	(11)	(19)	

13.

9	1	3	(13)
5	7	6	(18)
2	4	8	(14)
(16)	(12)	(17)	

14.

9	7	1	(17)
6	2	4	(12)
3	5	8	(16)
(18)	(14)	(13)	

15.

7	4	9	(20)
1	6	3	(10)
5	8	2	(15)
(13)	(18)	(14)	

16.

3	2	4	(9)
9	7	5	(21)
6	1	8	(15)
(18)	(10)	(17)	

17.

1	8	5	(14)
6	2	7	(15)
4	9	3	(16)
(11)	(19)	(15)	

18.

7	6	4	(17)
1	3	2	(6)
8	5	9	(22)
(16)	(14)	(15)	

19.

1	2	3	(6)
9	5	6	(20)
8	4	7	(19)
(18)	(11)	(16)	

20.

2	6	5	(13)
9	1	4	(14)
8	7	3	(18)
(19)	(14)	(12)	

21.

2	1	3	(6)
5	4	6	(15)
8	7	9	(24)
(15)	(12)	(18)	

22.

5	1	6	(12)
2	7	3	(12)
8	9	4	(21)
(15)	(17)	(13)	

23.

8	6	3	(17)
9	2	5	(16)
1	4	7	(12)
(18)	(12)	(15)	

24.

4	1	5	(10)
2	6	7	(15)
3	9	8	(20)
(9)	(16)	(20)	

Answers

25.

1	9	4	(14)
2	8	5	(15)
3	7	6	(16)

(6) (24) (15)

26.

1	7	4	(12)
6	2	8	(16)
5	9	3	(17)

(12) (18) (15)

27.

8	2	9	(19)
3	5	7	(15)
6	4	1	(11)

(17) (11) (17)

28.

2	3	4	(9)
5	6	7	(18)
8	9	1	(18)

(15) (18) (12)

29.

5	6	7	(18)
8	9	1	(18)
2	3	4	(11)

(15) (18) (12)

30.

9	1	2	(12)
3	4	5	(12)
6	7	8	(21)

(18) (12) (15)

31.

1	8	2	(11)
5	7	6	(18)
3	9	4	(16)
(9)	(24)	(12)	

32.

2	7	4	(13)
6	1	9	(16)
5	8	3	(16)
(13)	(16)	(16)	

33.

8	4	6	(18)
1	5	3	(9)
7	9	2	(18)
(16)	(18)	(11)	

34.

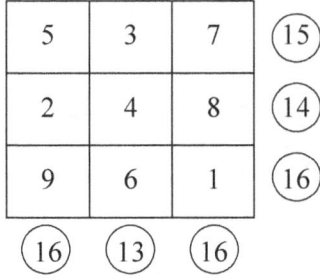

5	3	7	(15)
2	4	8	(14)
9	6	1	(16)
(16)	(13)	(16)	

35.

4	8	1	(13)
7	5	2	(14)
3	6	9	(18)
(14)	(19)	(12)	

36.

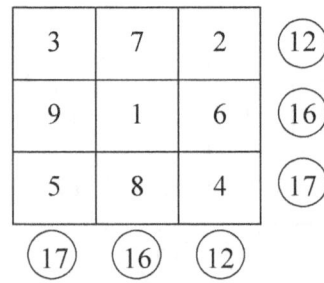

3	7	2	(12)
9	1	6	(16)
5	8	4	(17)
(17)	(16)	(12)	

37.

6	1	2	(9)
3	7	4	(14)
9	8	5	(22)

(18) (16) (11)

38.

5	4	1	(10)
8	6	3	(17)
9	2	7	(18)

(22) (12) (11)

39.

7	2	4	(13)
1	8	9	(18)
6	3	5	(14)

(14) (13) (18)

40.

4	7	9	(20)
1	3	5	(9)
2	6	8	(16)

(7) (16) (22)

41.

5	3	7	(15)
8	9	2	(19)
1	4	6	(11)

(14) (16) (15)

42.

5	4	7	(16)
6	1	2	(9)
9	3	8	(20)

(20) (8) (17)

43.

1	2	7	(10)
9	5	8	(22)
3	4	6	(13)
(13)	(11)	(21)	

44.

4	7	9	(20)
8	2	6	(16)
5	3	1	(9)
(17)	(12)	(16)	

45.

6	3	4	(13)
5	9	7	(21)
1	2	8	(11)
(12)	(14)	(19)	

46.

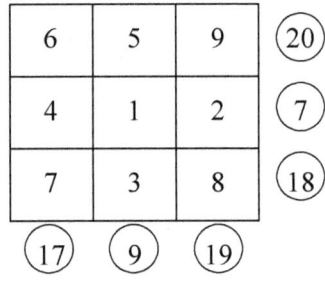

6	5	9	(20)
4	1	2	(7)
7	3	8	(18)
(17)	(9)	(19)	

47.

1	3	5	(9)
2	6	4	(12)
9	7	8	(24)
(12)	(16)	(17)	

48.

5	9	4	(18)
1	3	8	(12)
6	7	2	(15)
(12)	(19)	(14)	

49.

5	3	8	(16)
2	7	6	(15)
9	1	4	(17)
(16)	(11)	(18)	

50.

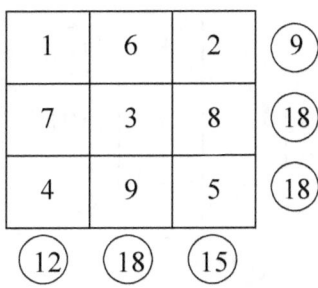

5	3	9	(17)
2	1	6	(9)
7	8	4	(19)
(14)	(12)	(19)	

51.

3	1	5	(9)
7	2	4	(13)
9	8	6	(23)
(19)	(11)	(15)	

52.

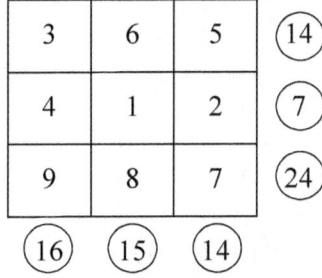

1	6	2	(9)
7	3	8	(18)
4	9	5	(18)
(12)	(18)	(15)	

53.

5	6	7	(18)
4	9	8	(21)
3	2	1	(6)
(12)	(17)	(16)	

54.

3	6	5	(14)
4	1	2	(7)
9	8	7	(24)
(16)	(15)	(14)	

Answers

55.

1	9	2	(12)
7	3	6	(16)
8	5	4	(17)
(16)	(17)	(12)	

56.

6	7	8	(21)
9	1	2	(12)
3	4	5	(12)
(18)	(12)	(15)	

57.

8	1	9	(18)
3	5	2	(10)
7	4	6	(17)
(18)	(10)	(17)	

58.

6	8	3	(17)
2	7	1	(10)
4	9	5	(18)
(12)	(24)	(9)	

59.

5	3	8	(16)
4	9	1	(14)
6	2	7	(15)
(15)	(14)	(16)	

60.

1	5	4	(10)
6	9	8	(23)
2	7	3	(12)
(9)	(21)	(15)	

Answers

61.

2	3	7	(12)
4	5	9	(18)
8	6	1	(15)
(14)	(14)	(17)	

62.

2	6	8	(16)
4	7	1	(12)
5	3	9	(17)
(11)	(16)	(18)	

63.

1	9	3	(13)
6	5	7	(18)
4	8	2	(14)
(11)	(22)	(12)	

64.

3	4	2	(9)
7	5	9	(21)
1	6	8	(15)
(11)	(15)	(19)	

65.

8	9	6	(23)
1	7	3	(11)
4	5	2	(11)
(13)	(21)	(11)	

66.

9	6	5	(20)
4	1	3	(8)
7	2	8	(17)
(20)	(9)	(16)	

Answers

67.

4	1	7	(12)
2	5	3	(10)
6	9	8	(23)

(12) (15) (19)

68.

5	3	1	(9)
9	7	4	(20)
8	6	2	(16)

(22) (16) (7)

69.

7	9	6	(22)
8	3	1	(12)
4	5	2	(11)

(19) (17) (9)

70.

6	5	4	(15)
7	9	1	(17)
8	2	3	(13)

(21) (16) (8)

71.

3	6	1	(10)
9	5	8	(22)
2	7	4	(13)

(14) (18) (13)

72.

6	7	9	(22)
1	2	3	(6)
4	5	8	(17)

(11) (14) (20)

73.

6	9	4	(19)
1	3	8	(12)
5	2	7	(14)
(12)	(14)	(19)	

74.

5	6	4	(15)
9	1	8	(18)
2	7	3	(12)
(16)	(14)	(15)	

75.

6	2	4	(12)
9	8	1	(18)
3	7	5	(15)
(18)	(17)	(10)	

76.

9	7	3	(19)
1	4	2	(7)
5	8	6	(19)
(15)	(19)	(11)	

77.

1	6	7	(14)
2	5	8	(15)
3	4	9	(16)
(6)	(15)	(24)	

78.

9	8	2	(19)
7	6	5	(18)
4	1	3	(8)
(20)	(15)	(10)